NATURAL LAW OF PROGRESSION

Table of Contents

INTRODUCTION 2
DIALECTICAL MATERIALISM 4
MARXISM 7
DIALECTICS AND METAPHYSICS 10
MATERIALISM V/s IDEALISM 18
LAW OF DIALECTICAL MATERIALISM 25
   1.   Law of Transformation of Quantitative to Qualitative Change: 25
   2.   Law of Unity and Struggles of Opposites: 30
   3.   The Law of the Negation of Negation: 36
DIALECTICAL MATERIALISM AND REVOLUTION 40
MARX'S DIALETICAL MATERIALISM 48
MARX'S VIEW ON CLASS STRUGGLE 52
ENGELS'S DIALETICAL MATERIALISM 54
CHARLES' DARWIN'S DIALECTICAL MATERIALISM 58
TECHNOLOGICAL REVOLUTION IN THE PROCESS OF HUMAN EVOLUTION 61
SUMMARY 70

INTRODUCTION

The inquest of law for the progression of human being hasn't been yet fulfilled universally, although various theorists have provided a reasonable and verifiable explanation. Such as Darwin's law of evolution confirms its theory while analysing more latest artefacts and species found in recent time. Also. The Marxist outlook of Dialectical Materialism that examines the subjects of the world in relation to each other within a dynamic, evolutionary environment, in contrast to metaphysical materialism, which examines parts of the world within a static, isolated environment.

This dialectical method of thought, later extended to the phenomena of nature, developed into the dialectical method of apprehending nature. This regards the phenomena of nature as being in constant movement and undergoing constant change, and the development of nature as the result of the development of the contradictions in nature, as the result of the interaction of opposed forces in nature.

Dialectical materialism accepts the evolution of the natural world and the emergence of new qualities of being at new stages of evolution. However, both theories somehow contradicted on some point and failed in manifesting the right outcome.

This book will provide an in-depth analysis regarding the law of progression analysing the real cause of development based on acquired knowledge using existing materialist theories where applicable.

DIALECTICAL MATERIALISM

Dialectical materialism can be broken down into its respective components for a better understanding. Dialectics describes the scientific method Marxists use to analyze the world around them. Materialism represents Marxists' conception of the reality dialectics is intended to analyze.

Dialectics, as a method of analysis, takes into account the interconnectedness of nature, the contradictions and state of continuous change inherent in it, and the process by which natural quantitative change leads to qualitative change. Simply put, dialectics holds that all things are in a constant state of change, that this continual change is a result of interactions and conflicts, and that many small hidden changes add up until the thing in question has been qualitatively transformed into something different.

The process by which water is transformed into steam, heating it until it passes the boiling point, illustrates the concept of dialectics at work. This understanding of development and change refutes the argument that class society is based on natural human greed. The development of class society came from the material interactions and conflicts that humans have faced over history.

A belief in dialectical materialism does not validate the oppression and exploitation of the working masses within this development of class society. Marxists argue that this scientific view analyzes how humanity and society have developed so that it can be changed. Most importantly, it instills the knowledge of human agency in history that people are in fact able to change the oppressive society that they live in, and that society cannot possibly stay the same as the material world changes.

Materialism is the Marxist conception of nature as it exists without any supernatural or mystical dimension. Materialism holds that objective reality exists independent of human consciousness and that matter is primary.

Dialectical materialism shows that people's thoughts, characters and actions are shaped by the conditions in the world around them, the material world. When people look at the world through the lens of dialectical materialism they can see the logical development of beliefs and thoughts, actions and events, and even human history as a whole.

Dialectical materialism implies that capitalism, like everything else, has a birth, a development, and will have an end.

MARXISM

Marxism is the name given to the body of ideas first worked out by Karl Marx (1818-1883) and Friedrich Engels (1820-1895). It is a conflict theory with argues that there is conflict between different groups within society. The way society produces the things we need (mode of production) exploits the proletariat because the bourgeoisie (ruling class) benefit from the working class labor.

In their totality, these ideas provide a fully worked-out theoretical basis for the struggle of the working class to attain a higher form of human society – socialism.

The study of Marxism falls under three main headings, corresponding broadly to philosophy, social history and economics – Dialectical Materialism, Historical Materialism and Marxist Economics. These are the famous "Three component parts of Marxism".

Dialectical materialism is the world outlook of the Marxist-Leninist party. It is called dialectical materialism because its approach to the phenomena of nature, its method of studying and apprehending them, is dialectical, while its interpretation of the phenomena of nature, its conception of these phenomena, its theory, is materialistic.

Historical materialism is the extension of the principles of dialectical materialism to the study of social life, an application of the principles of dialectical materialism to the phenomena of the life of society, to the study of society and of its history.

Marxian economics focuses on the role of labor in the development of an economy. Marxian economics argues that the specialization of the labour force, coupled with a growing population, pushes wages down and that the value placed on goods and services does not accurately account for the true cost of labour.

DIALECTICS AND METAPHYSICS

The Marxist view of the world is not only materialist, but also dialectical. For its critics, the dialectic is portrayed as something totally mystical, and therefore irrelevant. But this is certainly not the case. The dialectical method is simply an attempt to understand more clearly our real interdependent world. Dialectics, states Engels, "is nothing more than the science of the general laws of motion and development of nature, human society and thought." Put simply, it is the logic of motion.

It is obvious to most people that we do not live in a static world. In fact, everything in nature is in a state of constant change. "Motion is the mode of existence of matter," states Engels. "Never anywhere has there been matter without motion, nor can there be." The earth revolves continually around its axis, and in turn, itself revolves around the sun. This results in the day and night, and the different seasons that we experience throughout the year. We are born, grow up, grow old and eventually die. Everything is moving, changing, either rising and developing or declining and dying away. Any equilibrium is only relative, and only has meaning in relation to other forms of motion.

"When we consider and reflect upon nature at large or the history of mankind or our own intellectual activity, at first we see the picture of an endless entanglement of relations and reactions, permutations and combinations, in which nothing remains what, where, and as it was, but everything moves, changes, comes into being, and passes away," remarks Engels. "We see, therefore, at first the picture as a whole, with its individual parts still more or less kept in the background; we observe the movements, transitions, connections rather than the things that move, combine, and are connected. This primitive, naïve but intrinsically correct conception of the world are that of ancient Greek philosophy, and was first clearly formulated by Heraclitus: everything is and is not, for everything is fluid, is constantly changing, constantly coming into being and passing away."

The Greeks made a whole series of revolutionary discoveries and advances in natural science. Anaximander made a map of the world, and wrote a book on cosmology, from which only a few fragments survive. The Antikythera mechanism, as it is called, appears to be the remains of a clockwork planetarium dating back to the first century BC.

Given the limited knowledge of the time, many were anticipations and inspired guesses. Under slave society, these brilliant inventions could not be put to productive use and were simply regarded as playthings for amusement. The real advances in natural science took place in the mid-fifteenth century.

The new methods of investigation meant the division of nature into its individual parts, allowing objects and processes to be classified. While this provided massive amount of data, objects were analyzed in isolation and not in their living environment. This produced a narrow, rigid, metaphysical mode of thought that has become the hallmark of empiricism. "The Facts" became the all important feature. Facts alone are wanted in life.

"To the metaphysician things and their mental reflexes, ideas, are isolated, are to be considered one after the other and apart from each other, are objects of investigation fixed, rigid, given once and for all", states Engels. "He thinks in absolutely irreconcilable antitheses.' His communication is "yea, yea; nay, nay"; for whatsoever is more than this cometh of evil.' For him a thing either exists or does not exist; a thing cannot at the same time be itself and something else. Positive and negative absolutely exclude one another; cause and effect stand in rigid antithesis one to another.

"At first sight, this mode of thinking seems to us very luminous, because it is that of so-called sound common sense. Only sound common sense, respectable fellow that he is, in the homely realm of his own four walls, has very wonderful adventures directly he ventures out into the wide world of research. And the metaphysical mode of thought, justifiable and necessary as it is in a number of domains whose extent varies according to the nature of the particular object of investigation, sooner or later reaches a limit beyond which it becomes one-sided, restricted, abstract, lost in insoluble contradictions. In the contemplation of individual things, it forgets the connection between them; in the contemplation of their existence it forgets the beginning and the end of that existence; of their repose, it forgets their motion. It cannot see the wood for the trees."

Engels goes on to explain that for everyday purposes we know whether an animal is alive or not. But upon closer examination, we are forced to recognize that is not a simple straightforward question. On the contrary, it is a complex question.

There are raging debates even today as to when life begins in the mothers' womb. Likewise, it is just as difficult to say when the exact moment of death occurs, as physiology proves that death is not a single instantaneous act, but a protracted process.

In the brilliant words of the Greek philosopher Heraclitus, "It is the same thing in us that is living and dead, asleep and awake, young and old; each changes place and becomes the other. We step and we do not step into the same stream; we are and we are not."

Not everything is as appears on the surface of things. Every species, every aspect of organic life, is every moment the same and not the same. It develops by assimilating matter from without and simultaneously discards other unwanted matter; continually some cells die, while others are renewed. Over time, the body is completely transformed, renewed from top to bottom. Therefore, every organic entity is both itself and yet something other than itself.

This phenomenon cannot be explained by metaphysical thought or formal logic. This approach is incapable of explaining contradiction. This contradictory reality does not enter the realm of common sense reasoning. Dialectics, on the other hand, comprehends things in their connection, development, and motion. As far as Engels was concerned, "Nature is the proof of dialectics."

MATERIALISM V/s IDEALISM

"The philosophy of Marxism is materialism", states Lenin. Philosophy itself fits into two great ideological camps: materialism and idealism. Before we proceed, even these terms need an explanation. To begin with, materialism and idealism have nothing whatsoever in common with their everyday usage, where materialism is associated with material greed and swindling (in short, the morality of present-day capitalism) and idealism with high ideals and virtue. Far from it!

Philosophical materialism is the outlook which explains that there is only one material world. There is no Heaven or Hell. The universe, which has always existed and is not the creation of any supernatural being, is in the process of constant flux. Human beings are a part of nature, and evolved from lower forms of life, whose origins sprung from a lifeless planet some 3.6 billion or so years ago.

With the evolution of life, at a certain stage, came the development of animals with a nervous system, and eventually human beings with a large brain. With humans emerged human thought and consciousness. The human brain alone is capable of producing general ideas, i.e., thinking. Therefore matter, which existed eternally, existed and still exists independently of the mind and human beings. Things existed long before any awareness of them arose or could have arisen on the part of living organisms.

For materialists there is no consciousness apart from the living brain, which is part of a material body. A mind without a body is an absurdity. Matter is not a product of mind, but mind itself is the highest product of matter. Ideas are simply a reflection of the independent material world that surrounds us. Things reflected in a mirror do not depend on this reflection for their existence. "All ideas are taken from experience, are reflections – true or distorted – of reality," states Engels. Or to use the words of Marx, "Life is not determined by consciousness, but consciousness by life."

Marxists do not deny that mind, consciousness, thought, will, feeling or sensation is real. What materialists deny is that the thing called "the mind" exists separately from the body. Mind is not distinct from the body. Thinking is the product of the brain which is the organ of thought.

Yet this does not mean that our consciousness is a lifeless mirror of nature. Human beings relate to their surroundings; they are aware of their surroundings and react accordingly; in turn, the environment reacts back upon them. While rooted in material conditions, human beings generalize and think creatively. They in turn change their material surroundings.

On the other hand, philosophical idealism states that the material world is not real but is simply the reflection of the world of ideas. There are different forms of idealism, but all essentially explain that ideas are primary and matter if it exists at all, secondary. For the idealists, ideas are dissevered from matter, from nature. This is Hegel's conception of the Absolute Idea or what amounts to God. Philosophical idealism opens the road, in one way or another, to the defense of or support for religion and superstition. Not only is this outlook false, it is also profoundly conservative, leading us to the pessimistic conclusion that we can never understand the "mysterious ways" of the world.

Whereas materialism understands that human beings not only observe the real world, but can change it, and in doing so, change themselves.

The idealist view of the world grew out of the division of labor between physical and mental labor. This division constituted an enormous advance as it freed a section of society from physical work and allowed them the time to develop science and technology. However, the further removed from physical labor, the more abstract became their ideas. And when thinkers separate their ideas from the real world, they become increasingly consumed by abstract "pure thought" and end up with all types of fantasies.

Today, cosmology is dominated by complex abstract mathematical conceptions, which have led to all sorts of weird and wonderful erroneous theories: the Big Bang, beginning of time, parallel universes, etc. Every break with practice leads to a one-sided idealism.

The materialist outlook has a long history stretching back to the ancient Greeks of Anaxagoras (c.500 – 428 BC) and Democritus (c.460 – c.370 BC). With the collapse of Ancient Greece, this rational outlook was cut across for a whole historical epoch, and only after the reawakening of thought following the demise of the Christian Middle Ages was there a revival of philosophy and natural science. From the seventeenth century, the home of modern materialism was England. "The real progenitor of English materialism is Bacon," states Marx.

The materialism of Francis Bacon (1561 – 1626) was then systemized and developed by Thomas Hobbes (1588 – 1679), whose ideas were, in turn, developed by John Locke (1632 – 1704). The latter already thought it possible that matter could posses the faculty of thinking. It is no accident that these advances in human thought coincided with the rise of the bourgeoisie and great advances in science, particularly mechanics, astronomy and medicine. These great thinkers, in turn, provided the breakthrough with the brilliant school of French materialists of the eighteenth century, most notably René Descartes (1596 – 1650).

It was their materialism and rationalism that became the creed of the Great French Revolution of 1789. These revolutionary thinkers recognized no external authority. Everything from religion to natural science, from society to political institutions, was subjected to the most searching criticism. Reason became the measure of everything.

LAW OF DIALECTICAL MATERIALISM

1. Law of Transformation of Quantitative to Qualitative Change:

"Dialectics, the most complete, comprehensive and profound theory of development, is the heart and soul of Marxism-Leninism, its theoretical foundation. The universal-laws of dialectics reveal the essential features of any developing phenomenon, no matter to what field of activity it may belong". The dialectical materialism does not treat the nature or universe as stable or immobile the concept of development lies in this notion.

There is a continuous struggle between the opposite forces and this struggle, according to Marx and Engels, is the key to all sorts of progress. Dialectical materialism further states that the change or evolution from lower to higher, from Quantitative to Qualitative is never slow or gradual or smooth. It is sudden or abrupt. The real development of society envisages such a transformation.

1. P. Sheptulin remarks "The totalities of properties that make a particular thing what it is, is called its Quality. The totality of properties indicating a thing's dimensions or magnitude is called its Quantity. Dialectical materialism is not content with asserting that everything develops. The development or transformation is from Quantity to Quality."

Now let us explain what is meant by Quantitative and Qualitative change. The first basic law of dialectical materialism is that transformation of change may be Quantitative or Qualitative. All change has a Quantitative aspect. That is there may be decrease or increase of the thing. But quantitative change decrease or increase cannot go indefinitely.

It has its limitation. After certain point the Quantitative change may turn into Qualitative change. When the water is being heated it becomes hotter and hotter and after some time the water is converted into vapor. This conversion of water into vapor is Qualitative change.

Similarly, when the water is cooled and the temperature is brought down to the freezing point the water becomes ice.

The water is converted into ice but both are not the same. Quantitative changes occur constantly and gradually.

What about Qualitative change? Qualitative changes in a thing are a result of accumulated quantitative changes in it. So Quantitative and Qualitative changes in a thing are, in a sense, matter of stages. After a particular stage Quantitative change does not occur. Again, Qualitative change assumes the form of leaps.

There is a break or discontinuity which is absent in Quantitative change. Leap is a form of development that occurs much Quicker than the continual development. Leap form of development is characterized by intensity. It is really a breakthrough. Quantitative change is gradual, Qualitative change is abrupt.

The law states that there is an interconnection and interaction between the quantitative and qualitative aspects of an object thanks to which small, at first imperceptible, quantitative changes, accumulating gradually, sooner or later upset the proportion of that object and evoke the fundamental qualitative changes which take place in the form of leaps and whose occurrence depends on the nature of the objects in question and the conditions of their development in diverse forms. Knowledge of this law is vital to the understanding of development.

It provides a guideline for examining and studying phenomena as the unity of their qualitative and quantitative aspects, for seeing the complex interconnections and interactions of these aspects, and the changes in the relationships between them.

Engels borrowed the concept of Quantitative-Qualitative change from science and applied it to society. With the rapid growth of industrial capitalism wealth in the form of money is accumulated in the hands of few and the number of property less proletariat begins to rise and this proceeds unabated. When enough people have proletarianised to make capitalism mature quantitative change gives rise to Qualitative change.

2. Law of Unity and Struggles of Opposites:

1. P. Sheptulin defines opposites and contradictions in the following words:

"Aspects in which changes move in opposite directions and which have opposite trends of functioning and development are called opposites, while the interaction of these aspects constitutes a contradiction".

Every phenomenon is characterized by certain opposites and contradictions. This is the property of the phenomenon. For example, in capitalist society there are two antagonistic classes bourgeoisie and proletariat.

The interests, objectives, attitudes of these two classes are diametrically opposite. But they exist side by side and this is due to the interdependence and interconnection and interpenetration of opposites.

The opposites have different aspects of functioning and development and have different directions of change. But in spite of this the opposites do not eliminate each other they co-exist in an unbreakable unity and interdependence. This is an interesting characteristic of all opposites. Let us illustrate our point. In all class societies, Marx and Engels have said, there are mainly two class's proletariats and bourgeoisie or capitalists.

There are conflicts and contradictions and in spite of this both the classes exist side by side and this coexistence is inevitable. One cannot exist without the other. But a situation arrives when the coexistence becomes absolutely impossible and this finally leads to revolution or class struggle.

It has been claimed by Marx that after the revolution the proletarian class will establish its supremacy and create a classless society which is called a communism. Whether a communist society will bring about an end of contradiction is a debatable issue. But Marx and Engels have explained the matter from the standpoint of historical materialism.

The law of dialectics states that the struggle of opposites cannot be underesti¬mated. Rather, it is the motive force of social development. Lenin once said, "Development is the struggle of opposites."

This development or motion is self-development or self-motion. That is, the development is resulting from the struggle of opposites is not caused by external forces. This motion is quite relevant to dialectical materialism. This principle of dialectics has its own laws of motion. This is to be carefully remembered.

The contradictions are not immobile or immutable. Once they have arisen they develop and pass through definite stages. For the disappearance and replacement of contradictions two conditions are to be fulfilled.

One is contradictions must be fully revealed, and the other is, they must be fully developed. When these two conditions are fulfilled a situation for the leap will emerge. The old phenomenon will disappear and will be replaced by a new one which will be qualitatively higher or better than the earlier.

There are two stages of this development. First is a contradiction will unfold them and then they will be resolved. The contradiction first appears in the form of difference. Then this deepens into manifest contradiction.

In order to maximize profit the capitalist develops his productive system. Wealth, in the form of money, is concentrated in the hands of the few. Workers are more and more proletarianised.

Contradictions deepen. Ultimately demand is raised for the replacement of private property by the socialist property. When is such a demand made? The contradictions will arrive at a critical stage and the struggle between the opposites will reach the ultimate point. This is the stage of resolution of contradictions. Dialectical materialism attaches a good deal of importance to the resolution of contradictions.

In order to explain the law of struggle of opposites Cornforth cites an example. A cord will break when excessive load is put upon it. The qualitative change takes place as a result of the opposition set up between the tensile strength of the cord and the pull of the load.

Another example is when spring wheat is transformed into winter wheat, this is result of the opposition between the plant's "conservatism" and the changing conditions of growth and development to which it is subjected; at a certain point the influence of the latter overcomes the former.

3. The Law of the Negation of Negation:

Marx has said "In no sphere can one undergo a development without negating one's previous mode of existence."

Negation is an inevitable and logical element of development. Marx and Engels have said that a very powerful precondition of social development is the negation of previous existence.

Now the question is what is negation?

"In ordinary consciousness the concept of negation is associated with the word "no", to negate to say "no" or to reject something." But dialectical materialism looks at the concept from different angle. Negation is an important element of progress. So it has deeper connotation. According to to Engels "Negation in dialectics does not mean simply saying no, or declaring that something does not exist, or destroying in the same way one likes."

In the opinion of A. P. Sheptulin "Dialectical negation is objective. It is the negation of one qualitative state and the formation of a new one. It stems from the development of the internal contradictions of a phenomenon and result from the struggle between internal opposite forces and tendencies; it is a connecting link between lower and higher".

Dialectical negation is an important factor of progress or development. This is a feature of dialectical negation. Another feature is it combines old and new. That is it is a connection between the two. The negation of the old force by the new removes the obstacle on the way of development. Once development appears the new force does not stand disconnected with the old force. In this way a chain of connecting continues to exist.

The negation carries with it the potentialities of new force. Otherwise the negation is meaningless. It performs the function because it has the ability to create something new. Lenin state that "Not empty negation, not futile negation, not sceptical negation, vacillation and doubt is characteristic and essential in dialectics which undoubtedly contains the elements of negation and indeed as its most important element no, but negation as a movement of development retaining the positive".

The law of negation of negation is a law whose operation conditions the connection and continuity between that which is negated and that which negates. For this reason dialectical negation is not naked, "needless" negation, rejecting all previous development, but the condition of development that retains and preserves in itself all progressive content of previous stages, repeats at a higher level certain features of initial stages and has in general, a progressive and ascending character.

DIALECTICAL MATERIALISM AND REVOLUTION

There are a variety of theories that have been proposed to explain the changing world as we know it. Various thinkers explain the different ways the world has been transformed to be a better place to live.

According to Karl Marx, 'The philosophers have only interpreted the world, in various ways. The point, however, is to change it." This brings to a new perception the need to transform the interpretation of the world. Marx goes on to assert that for the purpose of interpreting the world, there is the need for a scientific understanding of it so as to change it.

Consequently, Marx and his long-term Colleague Engels were looking for a scientific explanation that could candidly interpret human history and ultimate progress.

It is through such interest that Karl Marx and his colleague Frederick Engels came up with Dialectical Materialism, a derivative of the Hegelian dialectic, meant to comprehend the human history with a materialistic conception. While Marx credits Hegel for presenting dialectic, he criticizes him for transforming dialectics upside down. Marx asserts, "With Hegel, it is standing on its head. It must be turned right side up again. As explained by Marx, Hegelian dialectics lays prominence on ideas, it caters for the process of the human brain which is dialectical idealism and hence wrong.

Instead, Marx proposes that "the material world, is the world of production and other economic activity and so should deal not with the mental world of ideas'. This can be loosely interpreted to imply that Marx ignores the abstract and proposes the need for the concrete in interpreting the world. Marx refutes Hegel's dialectics as "product of the human mind which failed to interpret the material world."

The manner in which Marx understands the world is founded on production relation, modes of production, the techniques in which societies are structured to exploit the existing technological powers to effectively interact with their material surroundings. Consequently, Marx's dialectic approach is materialistic and is popularly referred to as "Dialectic Materialism." His understanding of human history is materialistic and focuses on human societies and how they developed with time; hence the term historical materialism.

Marx and Engels appropriately applied Hegel's dialectic to comprehend the human history, but they failed to apply it efficiently and hence came up with conflicting outcomes to my perception. To illustrate this, it is important to explore Fredrick Engel, and Lenin's descriptions in his "Science of Logic", Engel described materialistic dialectic and drew three laws of dialectics which include the following:

1. The law of the unity and conflict of opposites.

2. The law of the passage of quantitative changes into qualitative changes.

3. The law of the negation of the negation.

The first law of Unity and conflict of opposites is quite logical. It can be perceived that it is equally prudent to have unity and at the same time a conflict of interests. In this, Marx and Engel unwittingly or wittingly propose for the unity within weak or oppressed class replacing the unity of both opposite classes. They put the unity of opposites on various and opposing dimensions and making them rivals. Despite this, as pointed out by the law, they should be complementary. Having opposing characteristics, or being in different social classes do not necessarily imply enmity.

In his essay "On the Question of Dialectics", Lenin argues that "Development is the 'struggle' of opposites." Lenin goes on to state that "The unity (coincidence, identity, equal action) of opposites is conditional, temporary, transitory, relative. The struggle of mutually exclusive opposites is absolute, just as development and motion are absolute." While Lenin's argument appears quite plausible; he describes half-truth the law of dialectic as the unity of opposites; and when he talks about proletarian revolution, he ignores the unity of opposites remained in two opposite classes; the bourgeoisie and the proletariat. Instead; he explores the unity within the proletariat class.

The question at this instance can be positioned, where is the unity of opposites as a law of dialectics drew by Engels?

Another great communist leader, Mao criticised Stalin and the Soviet party for allowing the union to drift into a state whereby institutions quickly created an atmosphere to bring societal resources under the communist control and acknowledged as having permanent and universal validity. Mao is of the opinion that this had the capacity of suppressing political activity. He insisted that dialectical materialism was equally applicable to socialist society as it is to the capitalist phase. In his speech, Mao explores such an idea and pushes for full participation in the process of development as a cultural revolution. For Mao, the notion of continuous revolution meant that the purpose of the Communist Party was not to simply staff an authoritarian bureaucracy but to enable and guarantee a process of development which gave it the form of Marxism to popular aspirations and to investigate a progressive process of transformation.

Here, Mao's interpretation of dialectic materialism at the time of the Soviet Union between Soviet workers and peasants is different.

Mao writes ' Even under the social conditions existing in the Soviet Union, there is a difference between workers and peasants and this very difference is a contradiction, although, unlike the contradiction between labour and capital, it will not become intensified into antagonism or assume the form of class struggle; the workers and the peasants have established a firm alliance in the course of socialist construction and are gradually resolving this contradiction in the course of the advance from socialism to communism.'

Mao also politicises his ideology, when he talks about the socialist state, he proposes firm alliance between classes and believes in gradually resolving the contradiction in the course of development in the socialist state, while his position, in development of socialism from capitalism is different, and talks about not revolution. He says 'A revolution is not a dinner party, or writing an essay, or painting a picture, or doing embroidery; it cannot be so refined, so leisurely and gentle, so temperate, kind, courteous, restrained and magnanimous. A revolution is an insurrection, an act of violence by which one class overthrows another.'

MARX'S DIALETICAL MATERIALISM

Marx's own writings are almost exclusively concerned with understanding human history in terms of systemic processes, based on modes of production (broadly speaking, the ways in which societies are organized to employ their technological powers to interact with their material surroundings). This is called historical materialism. More narrowly, within the framework of this general theory of history, most of Marx's writing is devoted to an analysis of the specific structure and development of the capitalist economy.

The concept of dialectical materialism emerges from statements by Marx postface to his magnum opus, Capital. There he intends to use Hegelian dialectics but in revised form. He defends Hegel against those who view him as a "dead dog" and then says, "I openly avowed myself as the pupil of that mighty thinker Hegel." Marx credits Hegel with "being the first to present [dialectic's] form of working in a comprehensive and conscious manner". But he then criticizes Hegel for turning dialectics upside down: "With him it is standing on its head. It must be turned right side up again, if you would discover the rational kernel within the mystical shell."

Marx's criticism of Hegel asserts that Hegel's dialectics go astray by dealing with ideas, with the human mind. Hegel's dialectic inappropriately concerns "the process of the human brain"; it focuses on ideas. Hegel's thought is in fact sometimes called dialectical idealism, and Hegel himself is counted among a number of other philosophers known as the German idealists. Marx, on the contrary, believed that dialectics should deal not with the mental world of ideas but with "the material world", the world of production and other economic activity.

For Marx, human history cannot be fitted into any neat a priori schema. He explicitly rejects the idea of Hegel's followers that history can be understood as "a person apart, a metaphysical subject of which real human individuals are but the bearers". To interpret history as though previous social formations have somehow been aiming themselves toward the present state of affairs is "to misunderstand the historical movement by which the successive generations transformed the results acquired by the generations that preceded them". Marx's rejection of this sort of teleology was one reason for his enthusiastic (though not entirely uncritical) reception of Darwin's theory of natural selection.

According to Marx, dialectics is not a formula for generating predetermined outcomes but is a method for the empirical study of social processes in terms of interrelations, development, and transformation.

Despite Marx's insistence that humans are natural beings in an evolving, mutual relationship with the rest of nature, Marx's own writings pay inadequate attention to the ways in which human agency is constrained by such factors as biology, geography, and ecology.

MARX'S VIEW ON CLASS STRUGGLE

Class struggle is elucidated as the tension or antagonism which exists in society due to competing socioeconomic interests and desires between people of different classes. It is the main work of Marxian political philosophy. Marx wrote in The Communist Manifesto, "The history of all hitherto existing society is the history of class struggles." Class struggle pressed society from one stage to the next, in a dialectical process. In each stage, an ownership class controls the means of production while a lower class provides labour for production. The two classes come into conflict and that conflict leads to social change.

Marx stated that class conflict is the real dynamic force of human history. In Communist Manifesto (1848), Marx and Engels wrote that "The history of all hitherto existing society is the history of class struggles". In the capitalist societies, class differentiation is most clear, class consciousness is more developed and class conflict is more acute. Therefore, capitalism is the concluding point in the historical feature of bourgeois period. Society as a whole is more and more splitting up into two great hostile camps, into two great classes directly falling each other – bourgeoisie and proletariat. It can be established that

According to Marx, Class is rooted in social relations of production, and cannot be mentioned in the first place to relations of distribution and consumption or their ideological reflections.

Engels postulated three laws of dialectics from Hegel's Science of Logic. Engels elucidated these laws as the materialist dialectic in his work Dialectics of Nature:

The law of the unity and conflict of opposites

The law of the passage of quantitative changes into qualitative changes

The law of the negation of the negation

The first law, which originates with the ancient Ionian philosopher Heraclitus, was seen by both Hegel and Vladimir Lenin as the central feature of a dialectical understanding of things.

It is in this dialectic as it is here understood, that is, in the grasping of oppositions in their unity, or of the positive in the negative, that speculative thought consists. It is the most important aspect of dialectic.

The splitting of a single whole and the cognition of its contradictory parts is the essence (one of the "essentials", one of the principal, if not the principal, characteristics or features) of dialectics. That is precisely how Hegel, too, puts the matter.

The second law Hegel took from Ancient Greek philosophers, notably the paradox of the heap, and explanation by Aristotle and it is equated with what scientists call phase transitions. It may be traced to the ancient Ionian philosophers, particularly Anaximenes from whom Aristotle, Hegel, and Engels inherited the concept. For all these authors, one of the main illustrations is the phase transitions of water. There has also been an effort to apply this mechanism to social phenomena, whereby population increases result in changes in social structure. The law of the passage of quantitative changes into

Qualitative changes can also be applied to the process of social change and class conflict.

The third law, "negation of the negation", originated with Hegel. Although Hegel coined the term "negation of the negation", it gained its fame from Marx's using it in Capital. There Marx wrote this: "The [death] knell of capitalist private property sounds. The expropriators [capitalists] are expropriated. The capitalist mode of appropriation, the result of the capitalist mode of production, produces capitalist private property. This is the first negation [antithesis] of individual private property. [The "first negation", or antithesis, negates the thesis, which in this instance is feudalism, the economic system that preceded capitalism.] … But capitalist production begets, with the inexorability of a law of Nature, its own negation. It is the negation of the negation.

"Engels made constant use of the metaphysical insight that the higher level of existence emerges from and has its roots in the lower; that the higher level constitutes a new order of being with its irreducible laws; and that this process of evolutionary advance is governed by laws of development which reflect basic properties of 'matter in motion as a whole'."

CHARLES' DARWIN'S DIALECTICAL MATERIALISM

There is no direct evidence that Darwin wittingly went on to support the dialectical method, he approached his world's perception from a materialist standpoint. In attempting to explain what he perceived of the actual world, he made keen observations of the natural phenomena in the context of its surroundings, the process of their development and ultimate change. His observations were made in a diligent manner over a wide period and in a wide extent of phenomena. Darwin perceived the world dialectically; and this prompted him to take a bold, informed step that had never been taken before. One intriguing aspect of Darwin's interpretation is the amount of evidence he managed to gather in the process of substantiating his arguments.

One major argument of Darwin is that Evolution led to the "survival of the fittest", this concept serves to emanate another concept, the concept of "struggle for existence". Darwin acknowledges reading on Malthus on human populations and this enhanced his thinking of how selection exists in organisms. It is quite evident to me that Darwin never perceived competition among people to be an impetus to the struggle for existence; he rather perceived the struggle for existence as being the fight or rather struggle to survive in the entire environment; and this includes the struggle with other species and with the physical conditions of their environment.

While Engel had little time for Darwin's perception on Malthus; he is quick to point out that Darwin's theory of natural selection does not rely on Malthus. He points out that Darwin's theory relied on his gathered evidence rather than on Malthus readings. Engel asserts that if Malthus was wrong, it does not disqualify Darwin's arguments

Simply put in summary, all the discussed interpretations can be quite conflicting. It can, however, be derived that they all half-quoted the dialectic materialism as a conflict between opposites unwittingly or wittingly ignoring its prudence and vital aspect of the Unity of opposites. Consequently, this serves as a prompt to emerge with an alternative interpretation of human societies, the concept of "Technological Revolution in the Process of Human "Evolution"

TECHNOLOGICAL REVOLUTION IN THE PROCESS OF HUMAN EVOLUTION

The history of man is full of narratives of struggles, inventions and ultimate transformations. However, the main impetus to human revolution; hence evolution is technology. According to the Merriam Webster Dictionary, technology is "the application of scientific knowledge for practical purposes". While it can be argued that there was no science in the prehistoric period, it is of the perception that science is inherent to man. To begin with, the history of tools in the process of exploiting the available natural resources by man.

According to existing evidence, the earliest stone tool making was developed at least 2.6 million years ago. But why tools? The answer to this might include the fact that food resources grew scarce. This could be as a result of droughts, and extreme competition. Man who was exclusively a gatherer of readily available meals such as fruits and leaves could not come across them. There was intense competition, and man opted for an alternative solution to meals; then he came across nuts such as almonds, coconuts and others that needed to be broken. It might have been through this that man discovered that stones could be used to crash and get the meal within. Man not only used stone tools to crush nuts but extended their usage to include security.

Currently, the oldest stone tools, referred to as Oldowan toolkit is made up of hammer stones, stone cores, and sharp stone flakes. All these tools depict that they were used crash, and pierce. During this prehistoric period of human evolution; those who could efficiently use the stone tools became the ultimate survivors; superior class of people. Those who could not became the lesser beings.

The exploitation of tools extended from for food to for protection and hunting and this is where the discovery of stone tips, arrows and bows come in. According to archaeologists, the stone tips are among the earliest forms of weapons used by early man. It is recorded that the earliest surviving stone tips which have animal blood dates back almost 64,000 years ago in Natal, current South Africa. When man learned the use of stones to crush nuts, he realized that they could be exploited further, sharpened, put at the end of a stick and hence as an arrow.

Other discoveries such as the fire are also consequential. Meat, which was not easy to eat was roasted and eaten by man. Man's capacity to control fire among the early human's was indeed a turning point in their history. The fire was advantageous as it availed a source of warmth, protection, a technique for cooking and another improvement in the hunting experience of man.

There are several evidence which is conclusive and dating back to 300,000 years ago whereby flint blades were found at proximity with fossils. Prehistoric man was a social being, and they lived in groups. However, the discovery of the tools for hunting and fire served to transform the society at large, those who could effectively utilize the tools and fire emerged as superior. They could fight better, hunt better and eat better. Those who were not ingenious enough to utilize such discoveries lagged in the society and transformed to be dependent on the others. This might have led to a form of social inequality in the prehistoric period.

As human beings went through the process of revolution as a result of their various forms of inventions. The notion of living in a group that has mutual understanding and dependency became practical. This led to the emergence of communities that merged small and initially isolated groups. The conjoined communities emerged into societies that are currently referred to as ancient civilizations. Examples of known ancient civilizations include The Incas Civilization, The Aztecs Civilization, The Roma Civilization, The Persian Civilization, The Ancient Greek Civilization, The Chinese Civilization, The Mayan Civilization, The Indus Valley Civilization, The Mesopotamian Civilization, and the Egyptian Civilization among others that might not have been documented or discovered yet.

What makes such ancient civilization popular up to this time, and what might have made them famous and hence distinct from other societies during their time is technological advancement. For instance, Ancient Egyptian Civilization is popular with its Pyramids that were constructed using advanced technology. The Shadoof method of irrigation ensured that farming was efficient, the civilization had a high supply of both water and food; hence making it unique. Other nearby societies hence proved inferior to Egypt as a result of its advancement technologically. Up to now, scientists marvel at the wonder of the pyramids and how people during those dark times managed to come up with such sophisticated structures.

Ancient Mesopotamia is also currently perceived as a cradle of human discoveries. The invention of the wheel was in Mesopotamia. It was not used for transportation but rather as porter's wheels. The wheel further transformed at it could be used for irrigation, transport, and in pottery making. During such ancient times, the wheel propelled social inequalities; the rich could afford the wheels for transportation. This is quite similar to the current times, when the wheels, in the form of cars, can be afforded by well-to-do individuals in the society. All in all, such a discovery has highly transformed society. Other inventions in ancient Mesopotamia include the chariot, the sailboat, the plow, time, astronomy and astrology, the map, mathematics, and urban civilizations.

Indeed, Civilizations have come and gone; and how they diminished depicts the important role of technology. Man has experienced eras of wars that have wiped down civilizations and led to the emergence of newer civilization. For instance, in the two world wars fought, the concern was to ascertain superiority. The superiority in this regards how powerful a nation is as it pertains to the weaponry at its disposal. When the US joined the Second World War, it emerged as a superpower and this is because of its superior weaponry and fighting technique. This set it apart and created a tension with the Soviet Union during the time, culminating to the emergence of the Cold War. Man has been on the rush to come up with superior weaponry; hence the emergence of increasingly dangerous and sophisticated weapons such as chemical weapons and the nuclear weapons.

For instance, what is it that makes the US different from a nation like Uganda in Africa? Is it the people or the technology that is being utilized? America is superior because of its superior weaponry; speak of the nuclear weapons and nations such as North Korea, Iran, Iraq, Russia, and the UK will emerge.

The social tension that exists worldwide is a consequence of the extreme technological advancements. Man's capacity to utilize technology in the exploitation of his environment is what set him apart from others who lack that ability. If what we currently refer to as Third World nations were to wholly embrace technology; and exploit aspects of their lives such as weaponry, medicine, urbanization and other prudent aspects; then the form of discrimination that exists in discerning them from "Developing nations", "Developed Nations" and "industrialized nations" would not exist.

Industrialization is a whole new form of social structuring. We categorized nations and places according to the manner in which technology has been embraced. For instance, what is the implication of Silicon Valley? This is where technology is increasingly utilized. Increasingly sophisticated companies are located in Silicon Valley and as a result; all people, or rather majority of those who work there are Rich and enjoy the benefits that come with it. Consequently, it is high time we embrace the term "Technological Revolution in the Process of Evolution". Human beings are different as it pertains to the technology they exploit and how they exploit it.

SUMMARY

Karl Marx and Friedrich Engels were searching for a method where they can interpret the human history and und understand the law of its development.

They used Hegelian dialectics but in revised form. He gives credit to Hegel for his dialectic but he then criticises Hegel for turning dialectics upside down: "With him it is standing on its head. It must be turned right side up again, if you would discover the rational kernel within the mystical shell."

Marx says, Hegel's dialectics is in fact dialectical idealism. Marx, on the contrary, believed that dialectics should deal not with the mental world of ideas but with "the material world", the world of production and other economic activity.

As Marx, dialectics is not a formula for generating predetermined outcomes but is a method for the empirical study of social processes in terms of interrelations, development, and transformation.

Despite Marx's insistence that humans are natural beings in an evolving, mutual relationship with the rest of nature, Marx's own writings pay inadequate attention to the ways in which human agency is constrained by such factors as biology, geography, and ecology.

As we noticed in the transition from Ape to Man, from the prehistoric period of human evolution; those who could efficiently use the stone tools became the ultimate survivors; superior class of people and those who couldn't utilise remained the lesser beings.

The discovery of the tools for hunting and fire served to transform the society at large, those who could effectively utilise the tools and fire emerged as superior. They could fight better, hunt better and eat better.

Those who were not ingenious enough to utilize such discoveries lagged in the society and transformed to be dependent on the others. This might have led to a form of social inequality in the prehistoric period.

From this, observation it would be summarised that it is the struggle of man with nature is the precursor of human evolution and not the other way like revolution.

There is no doubt saying, in any developmental process, unity of opposites forces play, a vital role while contradiction such as social inequality and social classification come at later stages as a contradiction, which is inherent in every development.

That means, human development is an evolutionary process and not a revolutionary, however evolutionary human character emerges in the due course of development in the result of social inequality.

Even Marx sees Darwin evolution as a materialistic, however, there is no direct evidence that Darwin wittingly went on to support the dialectical method, he approached his world's perception from a materialist standpoint.

His observations were made in a diligent manner over a wide period and in a wide extent of phenomena. Darwin perceived the world dialectically; and this prompted him to take a bold, informed step that had never been taken before. One intriguing aspect of Darwin's interpretation is the amount of evidence he managed to gather in the process of substantiating his arguments.

Darwin acknowledges reading on Malthus on human populations and this enhanced his thinking of how selection exists in organisms. It is quite evident to me that Darwin never perceived competition among people to be an impetus to the struggle for existence; he rather perceived the struggle for existence as being the fight or rather struggle to survive in the entire environment; and this includes the struggle with other species and with the physical conditions of their environment.

www.ingramcontent.com/pod-product-compliance
Lightning Source LLC
Chambersburg PA
CBHW030728180526
45157CB00008BA/3089